HACK THE UNVIERSE

Exploring and Exploiting the Architecture of a simulated Universe

James Brine

Deep Thought
Publishing

HACK THE UNIVERSE
Copyright © 2023 James Brine

Second Edition
First Published: September 2023
Production Reference: 13371337
ISBN: 9798863845418

www.jamesbrine.com.au

Little Max sits down at the kitchen counter to do his second-grade maths homework. Regardless of the advanced nature of the civilisation that Max lives in, his family are against the idea of directly injecting knowledge into her brain and so she learns the old fashion way, by doing, and with the help of his family's universe simulation quantum computing voice assistant, Quanto™.

Nobody's really sure exactly how it works, but the general principle is that is generates everything that could possibly be which would include the answer to your question, or in some cases many answers, and selects one to return as a result. The result isn't always logical, but given that an entire universe supports the answer, it's generally accepted that any weirdness that ensues must be your own lack of understanding of just how things work.

Hey, Quanto,™, what do you get when you multiply six by nine?
Hi Annie, sure I can work that out for you...

> **CREATE UNIVERSE**
> **GENERATE SEED VALUES**
> **INTELLIGENCE SIMULATED SUCCESSFULLY**
> **EXTRACTING SOLUTION TO QUESTION USING**
> **BLACK HOLES**
> **SHUTTING DOWN UNIVERSE**

The answer is, 42.
Thanks, Quanto™!
"Nobody is quite sure how ethical this is, simulations are just that, they can't feel anything, and they certainly wouldn't even know they were being simulated" – *that*, thought the sort of people who build these things, would be highly improbable.

Table of Contents

Introduction

The concept of a simulated universe is not a new one, as it has appeared in philosophical and religious writings for centuries. However, in recent years, the concept has gained traction due to advances in technology and the availability of powerful computers. The idea is that our entire universe could be simulated within some sort of virtual reality using a form of advanced computing, perhaps similar to the methods used to create video games and other digital realms.

Proponents of a simulated universe point to the possible advantages of such a system. For example, a simulated universe could potentially act as a way to explain some of the mysteries of our universe, such as the fine tuning of physical constants or the emergence of complex patterns from the primitive cosmic soup. Additionally, by being able to simulate the universe, we could explore new physics without risking consequences to the real universe.

II. Rationale for a Simulated Universe

There are several theories that have been proposed as potential rationale for why our universe could be a simulation. One of the most popular theories was proposed in 2003, by Nick Bostrom, a philosopher at the University of Oxford. Bostrom argued that it is statistically more likely that we are in a simulation than living in the physical world. He based this on the assumption that given the potential for advanced computing abilities in the future, it is likely that civilisations will create a large number of simulations for their own purposes. Bostrom's argument is based on the idea that with a sufficient number of simulations, the likelihood of us being in one of them is greater than the likelihood of existing in an un-simulated form.

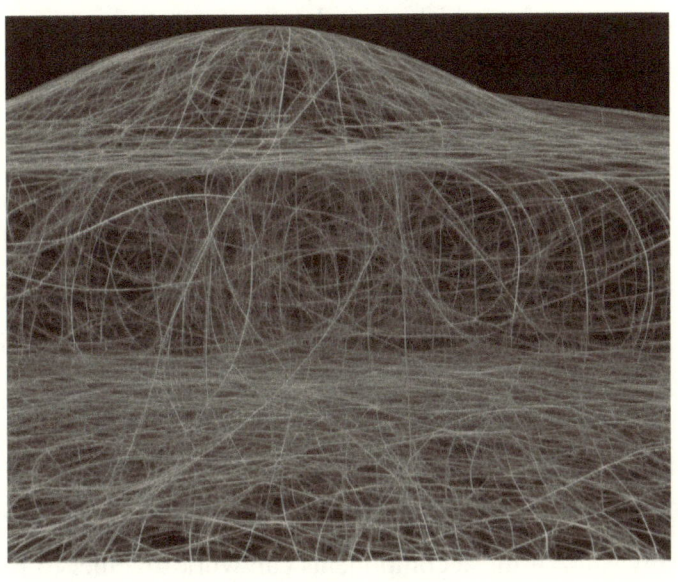

In addition to Bostrom's argument, recent scientific advances have also pointed to the possibility of a simulated universe. For example, quantum mechanics may play a role in the simulation of the universe, as suggested by a recent paper by García-Ripoll and Vidal from the Institute for Theoretical Physics in Spain. They suggested that quantum computing could be used to simulate physical systems at the level of individual particles. Similarly, the "holographic principle," which states that our universe is essentially two-dimensional, has been used to suggest that our universe could be simulated using a computer program.

III. Architecture of a Simulated Universe

In order to explore the potential architecture of a simulated universe, one must first understand the various components involved in such a universe. Processing power, data collection, monitoring, and garbage collection are all necessary components when considering the design of such a simulated universe. This chapter will explore how each of these components can work together to form a complete simulated universe.

Processing Power

Processing power in a simulated universe is provided by the creative forces within the universe. The ability to interpret and create new data is the first step in constructing a simulated universe. Through the creative forces of observation, shared model data, and quantum computing, a simulated universe is capable of forming and visualising complex systems.

In "Programming the Universe" by Seth Lloyd, he explains how quantum computing offers a far greater computing power than that of a traditional computer. This idea is based on the fact that some particles, called qubits, can exist in multiple states at the same time. This means that a quantum computer is able to perform calculations in parallel, allowing data to be processed much faster than traditional computers.

Additionally, the concept of string theory provides insight into how the creative forces of a simulated universe interact with the environment. One aspect of string theory is the notion of multiple dimensions, each dimension being a separate universe in its own right. In the simulated universe, these different dimensions can be used to create an array of possibilities from which a single, unified universe can be formed.

Data Collection

Data collection in a simulated universe is mainly accomplished through the use of black holes. Black holes act as a gateway to infinite universes, allowing the user to collect and analyse data from beyond our own realm. Additionally, black holes are thought to have the ability to draw out information from thoughts. This would be a perfect tool for a simulated universe, as it would enable users to extract and analyse data directly from their minds.

An alternative form of data collection is through the use of a different dimension. This dimension would allow the user to directly extract data from thought. This would enable a person or machine to access the thoughts and feelings of humans as a form of data collection.

Monitoring

In a simulated universe, monitoring is achieved by connecting directly to the senses or by using a debugger. A debugger is a powerful program used to monitor and analyse a system, and it can be used to monitor a simulated universe in much the same way. This would enable a user to track the performance of the simulated universe and identify any potential problems before they become an issue.

On the other hand, connecting directly to senses would give the user a better understanding of the simulated universe by allowing them to experience it firsthand. This would enable the user to not only track performance but also understand the environment and the motivations of characters within the simulated universe.

Garbage Collection

Garbage collection in a simulated universe is a process that eliminates any unnecessary data or objects that are no longer necessary or used. In order to do so, destructive forces such as black holes, supernovas, and other cosmic events would be used to "clean up" the environment. By eliminating these unnecessary objects, the simulated universe remains up to date and efficient.

Nested Realities: The Labyrinth of Simulated Universes

From the early days of philosophy, humanity has been captivated by the enigma of existence. As technology's footprint expanded, so did our scope of introspection. The once wild idea that perhaps our universe is a mere simulation has now rippled into mainstream discourse. But, to further complicate this web of reality, imagine a simulated universe inside another, much like the intricate layers of an onion. Science fiction, through myriad tales, has long danced around this concept. But what was once confined to the pages of fiction is now being taken seriously by some quarters of the scientific community. As we embark on this journey, it is essential to acknowledge the magnitude of complexities and the transformational ramifications of such a revelation.

Constraints of Nested Simulations

Every simulation, like a colossal machine, demands power. As we delve deeper into the nesting, each subsequent layer might require less computational detail, but the collective energy needs could be astronomical. The universe, in its grand design, must have a threshold. Just as an overloaded server struggles, the universe, too, may falter under the weight of excessive nested layers. But, there's an inherent paradox: while an infinite regress of simulations is enticing, each nested layer might find its computational capability diminished, setting a limit to the depth of this regression. For beings inside such a layered reality, these limitations would appear natural, a fundamental characteristic of their universe, unaware of the cascading dimensions above them.

Parallels with IT: Virtualisation & Containerisation

In the ever-evolving world of IT, the rise of virtualisation stands as a testament to human ingenuity. A single physical machine can host numerous virtual environments. This is reminiscent of the nested universes, each operating within parameters set by the superior layer. System administrators in IT deftly manage resources across these virtual machines, much like a grand cosmic orchestrator might ensure each nested universe gets its due resources. Each virtual environment is effectively insulated, suggesting that universes too might be shielded from one another, preventing a catastrophe in one from cascading downwards. And as efficiency remains a cornerstone in IT, with backup servers and redundancies, one wonders if the grand design of nested universes also incorporates such fail-safes.

The Observability Problem

Imagine a fish in a pond. Its world is bound by the water's edge, the universe beyond remaining largely elusive. Inhabitants of a nested universe could be similarly restricted, unable to perceive or comprehend the vastness that encircles them. However, the universe might offer subtle clues. Just as ripples on a pond's surface hint at external forces, certain cosmic events or phenomena might be echoes of the parent universe's workings. With technological strides, our ability to discern these 'echoes' could enhance, allowing glimpses into the true fabric of our reality.

Ethical and Philosophical Implications

The weight of creation is immeasurable. If indeed an advanced entity is behind our universe's design, the ethical responsibility is immense. Every joy, every sorrow, every dream experienced by countless beings within the nested universes arises from this grand act of creation. To intervene or not in the flow of these universes becomes an ethical quandary. Much like the philosophical discourses that surround deities and their role in human life, these questions take on a new meaning. The act of ending a simulation translates to annihilating an entire universe, with all its profound intricacies. And as we grapple with these thoughts, the question of consciousness takes centre stage. Do sentient beings in simulations possess rights? As artificial intelligence continues its relentless march, this question becomes ever more pertinent.

Conclusions and Future Prospects

As we stand on the precipice of understanding, the nested simulation theory extends an invitation to reconsider our perception of reality. This concept, though speculative now, challenges us to unify our scientific and philosophical pursuits. Our place in the potential hierarchy of universes awaits discovery, and as we refine our tools and perspectives, perhaps the layers of this cosmic onion will unfurl, revealing the truths we've so ardently sought.

IV. The Simulation Argument

Many scientists believe the universe is simulated, which is supported by researchers analysing mathematical. Nicholas Bostrom, a Swedish philosopher, proposed the Simulation Argument in 2003. Bostrom's argument is based on probability theory and suggests that at least one of the following statements is true:

Human civilisation or a comparable civilisation is unlikely to reach a level of technological maturity capable of producing simulated realities, or such simulations are physically impossible to construct.

- [ ] A comparable civilisation reaching aforementioned technological status will likely not produce a significant number of simulated realities (one that might push the probable existence of digital entities beyond the probable number of "real" entities in a Universe) for any of a number of reasons, such as diversion of computational processing power for other tasks, ethical considerations of holding entities captive in simulated realities, etc.

- "Any entities with our general set of experiences are almost certainly living in a simulation." ("Simulation hypothesis - Wikipedia")

- We are living in a reality in which post-humans have not developed yet and we are actually living in reality.

- We will have no way of knowing that we live in a simulation because we will never reach the technological capacity to realise the marks of a simulated reality.

Based on these premises, Bostrom points out that if technological perspectives grow exponentially, then eventually every possible universe would be able to be simulated. Therefore, the probability of us existing in a simulated universe is much greater than the probability that no simulation exists.

Quantum Computing

The potential for the simulated universe is further enhanced by the development and research of quantum computers. With the computing power provided by these machines, it is possible to simulate entire universes down to the subatomic level. Experiments conducted in the union of quantum physics and information theory have been able to simulate particles and states that occur in nature. This can be used to further prove the possibility that we are living in a simulated universe.

Another piece of evidence to a simulated universe is provided by quantum entanglement. Two particles that are entangled are linked in terms of their physical properties and are connected, regardless of how far apart they are. This phenomenon can be used to move information across universes, further implying the idea that the universe is simulated.

Potential for Existence

The possibility of a simulated universe has serious implications for the question of free will. If a simulation is run by a creator, it is unknown whether or not that creator has control over the simulated people- and if they have predetermined our lives or given us free will. It is still an unanswered question but may provide a new understanding of our existence.

The concept of a simulated universe also presents us with philosophical implications. If we exist in a simulated universe, then do our thoughts, aspiring and experiences have any real meaning? We simply may be existing in a program, where the people, events and moments have been predetermined by a higher force. This bears serious implications when it comes to morality, ethics, and other similar subjects.

Ultimately, it is still uncertain as to whether or not we are existing in a simulated universe; however, there is a growing body of evidence that supports this premise. With the development of quantum computing, we will be able to further explore the concept of a simulated universe. Until then, we will have to wait for further scientific advancements and discoveries.

CHAPTER TWO
The Information Dimension

What is Information?

Have you ever gazed upon a night sky and felt overwhelmed by the sheer immensity of it all? Every star, every solar system, every galaxy, every local group, every supercluster, as well as every particle of dust, carries with it a history, a piece of information. But what if these particles weren't just passive bystanders? What if they were active contributors to the very essence of reality?

This leads us to the mind-bending contemplation: what if the universe, in its entirety, is not just an amalgamation of matter but a vast reservoir of information?

Many have conceived of information as mere data, often visualised as zeros and ones flowing through computer systems. But a deeper dive reveals a more profound truth. Instead of being just a by-product, information might be the very fabric that weaves our reality together, making it a fundamental construct, rather than an emergent one.

Information Theory and Its Philosophical Ramifications

Information theory, as developed by Claude Shannon, laid the groundwork for understanding the transmission of data, but it has also sparked philosophical inquiries.

The pivot from viewing information as mere data to seeing it as an essential building block was heralded by Claude Shannon's seminal work, "The Mathematical Theory of Communication." Through this, he introduced entropy, linking it to the information content of a message. It wasn't just about transmitting messages anymore; it was about understanding the universe. Shannon's 1948 paper addressed the communication of digital data - or signals - over noisy channels. It looked at how to optimally encode signals for transmission, and with what measure of fidelity they should be received. The overarching goal of Shannon's paper was to provide a mathematical basis for communication and to quantify the efficiency of communication systems.

Information Theory states that when it comes to information, the universe is constantly evolving and in a continual state of change. Therefore, information is not static; instead, it is constantly changing. The theory argues that information is produced, consumed, and transformed when it interacts with other physical systems. Furthermore, it states that information exists within a hierarchy, with emergent, more complex information built on top of simpler information. This hierarchical relationship helps to explain why certain types of information are composed of others - for example, how a sentence is made up of words, which are the result of individual letters.

The implications of Shannon's ideas resonate beyond communication technology. If information is at the heart of reality, then every bit of knowledge, every shared idea or secret, becomes a part of the ever-evolving narrative of existence.

The Fluidity of Knowledge

Drawing parallels with Heraclitus's philosophy, where one cannot step into the same river twice, we're urged to see knowledge not as fixed but as constantly changing. In the realm of the Information Dimension, to hold onto rigid beliefs is to ignore the dynamic nature of reality.

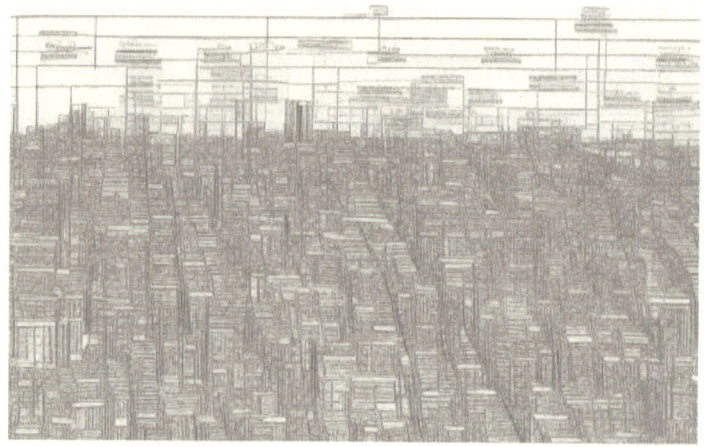

Modern educational structures often paint knowledge as static. Yet, if we accept the fluidity of knowledge in the Information Dimension, we're compelled to reconsider how we teach, learn, and grow. The unpredictability and limits of knowledge, as outlined by

Nassim Nicholas Taleb in "The Black Swan," emphasise that embracing uncertainty isn't just beneficial but essential.

Consciousness: The Master Information Processor

Our journey within the Information Dimension wouldn't be complete without delving into the role of consciousness. It's not just about perceiving information; it's about processing, understanding, and shaping it. Our thoughts, emotions, and experiences are not mere results of information processing; they actively contribute to the ongoing information stream. Consciousness is a complex and highly debated concept that is difficult to quantify. Generally speaking, consciousness can be thought of as an awareness of oneself, and the surroundings, that includes both physical and mental aspects. It can also be viewed as an experience of the self in relation to the outside world.

The presence and level of consciousness has been a matter of much debate among researchers, philosophers, and psychologists. It is hard to measure and define, but it is accepted that it is present in all living things, particularly humans. This can be evidenced by the presence of certain biological structures, such as the frontal lobe, which is believed to be involved with the awareness of thought. Humans are not alone in having the ability for consciousness, as there have been suggestions that the level of consciousness may extend to animals and even non-living things such as computer networks or artificial intelligence systems. Drawing upon the theories of Giulio Tononi and David Chalmers, one can argue that consciousness is more than just an emergent phenomenon. It is both a participant and a creator in the grand theatre of information.

How does our consciousness filter the plethora of data we encounter daily? Every sight, sound, and touch presents an information challenge, and yet, our minds seamlessly convert this barrage of data into coherent experiences. This transformation from raw information to personal understanding underscores the magical interplay between consciousness and information. Consciousness and the connection to the Information Dimension are discussed in greater detail in the following chapter "Consciousness Dimension"

The Interconnected Tapestry of Knowledge Transfer

Could the universe be a vast, interconnected network of information, where every piece of data, from the smallest subatomic particle to the grandest cosmic structure, contributes to the grand tapestry of existence? In the Information Dimension, no piece of information exists in isolation. There's a continuous dialogue – a dance – between individuals, societies, and even civilisations. This exchange goes beyond classrooms, books, or digital screens. It becomes a holistic system, interconnected and

evolving. Peter Russell's concept of the "global brain" offers a tantalising glimpse into this. As we become more interconnected, our collective consciousness, driven by technology, starts resembling a distributed cognitive system. Knowledge isn't just about passive accumulation; it's about active participation in the ever-unfolding story of the universe. Every exchange, every conversation, adds a new thread to the grand tapestry of existence.

External Collection and Storage Outside a Simulated Universe

In our journey into the Information Dimension, it's critical to entertain the possibility of external collection and storage mechanisms that lie outside the conventional confines of our perceived universe. One such proposition is that our universe, or perhaps multiverse, operates akin to a grand quantum computer. Each event, interaction, and thought contributes to a colossal computation—ultimately aimed at solving a complex problem or achieving a specific purpose.

Drawing parallels to the mechanics of a quantum computer, our universe might leverage superposition and entanglement to process vast amounts of information concurrently. The emergence of conscious beings, then, can be seen not as an evolutionary accident but as a design to boost the processing power of this universe-as-a-computer. Just as a computer stores data in its memory, the universe might have mechanisms to retain information externally, even beyond its apparent destruction.

In the realm of computing, hypervisors and containerisation offer us a metaphor that could illuminate our understanding of the simulated universe's architecture. When we think of a hypervisor, we imagine a platform that allows multiple virtual machines (VMs) to run concurrently, sharing the same physical resources but operating as if they were independent entities. Similarly, containers provide isolated environments for applications to run, abstracted from the underlying infrastructure.

These VMs and containers remain unaware of each other's specific operations, and more importantly, unaware of the overarching system controlling them—the hypervisor or the container orchestration platform.

Imagine our universe as one such container, operating within a vast infrastructure of multiple universes or realities.

Just as a containerised application is oblivious to the specifics of the external storage it's linked to, we, within our universe, might remain blind to vast reservoirs of external information or dimensions.

Symbolic links in computing allow different points in a file system to reference the same data. This means that diverse systems or applications can access the same data without duplicating it. Drawing a parallel, quantum entanglement can be viewed as a form of cosmic symbolic link. When two quantum particles become entangled, the state of one instantly affects the state of the other, no matter the distance between them, as if they are accessing the same shared data point. This phenomenon could be a hint of underlying shared storage or informational structure in the universe's fabric, unseen yet influential.

However, with knowledge comes curiosity and the inherent human drive to push boundaries. In the tech world, 'container escape' is a term used for exploits that allow an entity within a container to break out and access the broader system. Philosophically, if our universe is akin to a computational container, the concept of 'container escape' suggests a tantalising possibility: Could we find a way to peek outside our universe, to access or at least perceive the external storage or systems?

Such a quest to "break out" and glimpse beyond our container would be the ultimate test of the simulated universe theory. While the means to achieve this remain speculative, recognising parallels between our advancements in computational technologies and the nature of our universe could guide our investigations. Pursuing this line of inquiry would not only reshape our understanding of reality but might also provide methods to interact with, or at least understand, realms beyond our current perception.

The Role of Black Holes in Information Conservation

Black holes, enigmatic and captivating, have long been a topic of intrigue and speculation in both scientific and popular domains. At their core, black holes are regions of spacetime where gravity is so strong that nothing, not even light, can escape their grasp. This formidable force has profound implications on our understanding of the universe and the fabric of reality itself.

The birth of a black hole is typically tied to the life cycle of massive stars. When a star several times more massive than our sun exhausts its nuclear fuel, it undergoes a supernova explosion. Following this cataclysmic event, the core that remains can collapse under its gravitational pull, resulting in a singularity—a point in space where density becomes infinite. This singularity is enveloped by an event horizon, a boundary beyond which nothing can return. It is this region that we commonly refer to as the black hole.

A fascinating feature of black holes is their ability to warp time and space in their vicinity. As an object approaches a black hole, time for that object, from an external observer's perspective, seems to slow down. This effect, known as time dilation, is a direct result of Einstein's theory of general relativity. Interestingly, for someone approaching a black hole, they would perceive their own sense of time as normal but would see the universe outside accelerating. Recent theoretical physics has proposed many fascinating hypotheses regarding black holes. One prominent idea, stemming from the realms of quantum mechanics and general relativity, is the black hole information paradox. When matter falls into a black hole, the information associated with that matter seems to disappear, violating the principles of quantum mechanics that state information cannot be destroyed. Recent developments in theoretical physics suggest that black holes, long perceived as the ultimate devourers of information, might instead be fundamental in preserving it. The apparent paradox—that black holes both destroy and conserve information—can be reconciled if we view them as gateways or repositories to an external storage system outside our universe.

Stephen Hawking's idea that black holes emit radiation, now known as Hawking radiation, suggests that nothing is truly lost within a black hole. Instead, as the universe approaches its eventual fate—be it a Big Crunch, Heat Death, or otherwise—black holes might serve as the last bastions of information collection. Accumulating data and solutions achieved throughout the life of the universe, these enigmatic entities might eventually transfer this information to an external dimension, ensuring its conservation and potential application.

The Purpose of a Simulated Universe: Solving the Grand Problem

If we accept the notion of our universe as a simulated quantum computer, the subsequent question becomes: to what end? It's conceivable that this universe has been engineered to solve a problem so intricate and grand that it requires the entire breadth of cosmic history and the collective computational prowess of conscious beings to decipher. Each life lived, each civilisation risen and fallen, and every galaxy formed and destroyed might represent not mere historical events, but computational steps towards solving this grand problem. When the solution, or perhaps a set of solutions, is realised, the universe's primary purpose is achieved. The collected information, stored in black holes or other cosmic data banks, is then transferred externally preserving the insights gleaned from billions of years of computation.

CHAPTER THREE
The Consciousness Dimension

Consciousness as a Fundamental Dimension

As we traverse the vast landscape of the Consciousness Dimension, we're met with a paradigm-shifting perspective—one that places consciousness not as a mere outcome of our intricate neural connections but as a cornerstone of existence. The idea is both revolutionary and intriguing, pushing the boundaries of our accepted truths.

David Chalmers, in "The Conscious Mind," presented an enthralling exploration of consciousness, challenging traditional neuroscientific approaches. He posed the "hard problem of consciousness," emphasising the difficulty of explaining why and how we have subjective experiences. According to Chalmers, our cognitive functions, like learning or reasoning, can be broken down and understood through neurological processes. Yet, the subjective experience, the essence of consciousness, remains elusive.

The foundational nature of consciousness suggests that, rather than emerging from complex computations in the brain, consciousness might be a primary feature of the universe, akin to space and time. This view is resonant with Eastern philosophies, which have, for millennia, considered consciousness to be intrinsic to reality. The ancient Vedic texts, for instance, described the universe as a play of consciousness, where every element, living or inanimate, carries a spark of the divine.

Western philosophy too, albeit more sparsely, touched upon this theme. Plato's realm of ideals or forms postulates an existence beyond the tangible, one that is more real than our sensory experiences.

The Nature of Conscious Experience

As we delve deeper into the Consciousness Dimension, we encounter the intricacies of individual subjective experiences. Each person's reality is distinct, painted with the unique brushstrokes of their consciousness. Philosophers have long been fascinated by this aspect. Hegel's "Phenomenology of Spirit" explored the evolution of consciousness, tracing its path from immediate sensation to a state of self-awareness that recognises its connection with the universal.

This dialectical journey sees consciousness oscillate between opposites, constantly seeking synthesis and realisation. Moving from philosophical to practical realms, the question arises: What sets conscious entities apart from inanimate ones? Is it just complexity, or is there a deeper quality at play? Thomas Nagel's musings in "What Is It Like to Be a Bat?" shed light on the inherent challenge of understanding another's conscious experience. Every conscious entity, he suggests, possesses a unique subjective essence that cannot be fully grasped externally.

Interconnected Consciousness

The Consciousness Dimension suggests that while individual experiences are pronounced, they are not isolated. Our perceptions, emotions, and desires are intertwined in a vast cosmic tapestry. This sense of interconnected consciousness isn't novel. Carl Jung, with his concept of the collective unconscious, proposed the existence of a shared mental space housing universal archetypes and symbols that span cultures and epochs. This idea of inter-connectedness extends beyond the psychological domain. Quantum physics, with phenomena like entanglement, also hints at non-local connections, potentially suggesting a parallel with interconnected consciousness. Venturing further, Ervin Laszlo's "Science and the Akashic Field" postulates the existence of a cosmic memory bank—a field that connects all, pointing to a reality where everything is deeply interconnected.

The Transfer of Knowledge through Consciousness

In the Consciousness Dimension, the manner in which knowledge is acquired and shared transcends the conventional. Instead of being solely reliant on sensory or cognitive channels, it is inextricably tied to consciousness. Rupert Sheldrake's "The Presence of the Past" offers a glimpse into this perspective through the concept of morphic resonance, which postulates an inherent memory in nature. This memory allows patterns to repeat and knowledge to span spatial and temporal bounds. This view of knowledge transfer isn't just about data exchange; it's a dance of resonances, a harmonisation of consciousness.

Such a perspective suggests that rituals, symbols, and shared practices might be pathways to access collective memories—a vast repository of collective wisdom.

Consciousness in a Simulated Quantum Universe

In the realm of modern theoretical physics, the possibility of our universe being a grand simulation has gained traction. This idea, while seemingly far-fetched, finds grounding in the peculiarities of quantum mechanics. If we entertain the idea that our universe operates like a quantum computer, a new perspective on consciousness emerges. Within such a computational universe, consciousness is no mere passenger; it becomes an active participant, instrumental in the resolution of quantum states. Every conscious observation could potentially act as a 'quantum computation,' facilitating the realisation of one outcome over another.

Consciousness: The Quantum Resource

Considering the universe as a quantum simulator imparts a novel role to consciousness—as a vital resource. Quantum mechanics is imbued with intrinsic uncertainties.

However, the act of conscious observation collapses wave functions into definite states. In this framework, consciousness is not just a by-product; it's an essential tool, a resource, facilitating the computational processes of this quantum simulator. Without consciousness, the quantum universe might remain in a superposition, a haze of probabilities. Conscious entities become the driving force that refines these probabilities into the tangible reality we experience.

The Necessity of Consciousness for Information Collection

In a simulated quantum universe, information is paramount. If consciousness is integral to the extraction of quantum information, then its existence becomes not just incidental but necessary. This perspective postulates that for the quantum simulator to gather, process, and evolve information effectively, conscious entities are indispensable. It's akin to having sensors in a vast computational grid. Without these sensors (conscious entities), the system would lack the feedback mechanisms needed to process and refine its data.

Thus, consciousness becomes a prerequisite for the efficient functioning and evolution of this simulated quantum reality. These additional layers of understanding accentuate the complexity and centrality of consciousness, not just as an observer but as an active participant and even a requirement in the grand scheme of a quantum computational universe.

CHAPTER FOUR
The Perceptual Dimension

Perception as an Active Construct

Within the realm of multi-dimensional theories, the "perception dimension" emerges as a unique and intriguing proposition. Unlike spatial or temporal dimensions, the perception dimension doesn't exist in the tangible universe; rather, it embodies the collective imperfections, biases, and limitations inherent in the way sentient beings perceive reality. Each organism, based on its evolutionary path, sensory organs, and cognitive capabilities, sees, and interprets the world differently. These interpretations, limited and imperfect as they might be, consequently influence actions and decisions, thus playing a role in shaping the very reality we inhabit.

To understand the profound impact of the perception dimension, consider the butterfly effect. Even the tiniest of actions, stemming from flawed perceptions, can cascade into significant changes in the future. For instance, humans, with their binocular vision and limited colour perception, have built a world that reflects these limitations. If we perceived ultraviolet light or had a different field of view, our art, architecture, and even social dynamics would be vastly different. These alterations in reality are a direct consequence of our perceptions—or misperceptions.

When we extend this concept to a larger scale, encompassing all beings and their respective perceptions, we can appreciate the vast web of interactions and influences that continually mutate and transform our shared reality. A bat's echolocation or a mantis shrimp's incredible colour vision opens up worlds and realities that humans can't easily fathom. Their interactions with their environment, driven by these unique perceptions, inevitably cause ripples in the fabric of our shared reality.

Perception and Artificial Intelligence

The dynamics of the perception dimension take on a new dimension in the context of artificial intelligence, especially with image generation algorithms. AI does not "perceive" in the organic, sentient sense, but it processes and interprets data based on its programming and the datasets it's been trained on. Given a certain dataset, an AI can produce an infinite array of outputs, each slightly different from the others. However, things take a fascinating turn when AI starts feeding on its own outputs—a phenomenon called Model Autophagy Disorder (MAD).

MAD presents a peculiar scenario. When an AI continually processes its generated outputs, it's akin to a feedback loop. Initially, the outputs might show increasing randomness, novelty, or unpredictability, reflecting the AI's interpretation of its interpretation. But over time, two dominant patterns emerge. Either the AI starts producing increasingly similar, almost identical results, essentially getting stuck in a pattern, or it begins to deviate significantly from the original model, losing the essence of the original data.

The convergence towards similarity can be likened to an echo chamber effect. The AI, constantly recycling and reprocessing its own output, becomes trapped in a narrow perception band, much like humans who only interact with like-minded individuals and reinforce their own beliefs. The outputs become monotonous and lack diversity, a stark reminder of the importance of varied inputs and experiences in any learning system.

On the other hand, when the AI strays far from the original model, it reflects the malleability of reality under the influence of the perception dimension. The iterative mutations and changes accumulate to a point where the resultant output bears little resemblance to the original, much like a game of "Chinese whispers". This divergence underscores the unpredictable and transformative power of perception and interpretation, even in a system devoid of consciousness.

The Influence of Subjectivity on Perception

In the vast expanse of the Perceptual Dimension, perception emerges not merely as an observer but as an active participant in shaping reality. Here, the static, passive notions of perception, often taken for granted, dissolve into more fluid, dynamic interpretations. The philosophy of phenomenology, particularly Edmund Husserl's work "Ideas: General Introduction to Pure Phenomenology," beckons us toward this deeper understanding. Husserl dared to delve into consciousness, elucidating its active and intentional nature. Through his lens, perception isn't just a mirror reflecting reality; it's a creative force, actively constructing the realms we inhabit.

Venturing further into this dimension, we're confronted with the irrefutable power of subjectivity. It becomes evident that our perceptions are neither uniform nor universal, but intimately coloured by individual experiences. Maurice Merleau-Ponty, in his seminal work "Phenomenology of Perception," championed this notion. He elucidated how our subjective engagements with the world intricately interweave with our perceptions, making the two inseparable. Here, in the Perceptual Dimension, beliefs, cultural imprints, and unique personal experiences rise to prominence, underlining their decisive role in crafting the tapestry of our perceived universe.

Altered States of Perception

The dimension further surprises with its embrace of altered states of perception. Rather than sidelining them as anomalies or aberrations, it holds them in reverence, acknowledging their profound capacity to unveil alternate facets of reality. The scientific world, too, has been intrigued by such states, with psychologists and neuroscientists delving deep into the altered realms invoked by meditation, psychedelics, and more. Aldous Huxley, in his evocative narrative "The Doors of Perception," took readers on a journey through the altered terrains he traversed under the influence of mescaline.

Such firsthand accounts, coupled with empirical studies, prompt contemplation. Can these altered perceptions, by shattering conventional moulds, provide fresh, transformative insights?

Variability of Reality within the Dimension

Reality, within this dimension, emerges as a dynamic entity, continually reshaped by myriad perceptions. This fluidity resonates with constructivist theories that dwell upon the idea that realities are co-crafted - sculpted simultaneously by individuals and their surrounding societal ecosystems. Peter L. Berger and Thomas Luckmann, in their thought-provoking exploration "The Social Construction of Reality," delve into this dance of creation and sustenance, revealing the mechanisms by which humans collectively conjure and uphold their realities. Here, within the dimension's confines, it becomes abundantly clear that reality is not a monolithic entity but a mosaic, with each piece reflecting diverse interpretations borne from the subjectivity of perception.

As we retreat from the Perceptual Dimension, we're bequeathed with transformative insights. The journey compels a re-evaluation of our relationship with reality, urging acknowledgment of perception's active, pivotal role.

With every sensory experience, from the gentlest whisper to the most vivid hue, we're reminded of its power to mould our understanding.

As the layers unravel, the revelation that perception is not a mere passive reception, but an intricate act of creation becomes undeniable. Every belief held, every cultural story absorbed, every personal memory cherished, converges, influencing this ceaseless act of perception, perpetually constructing our realities. Amidst this vast landscape, the phenomenon of altered states stands as a beacon, illuminating the myriad possibilities that lie beyond conventional perception. Such states, whether invoked by the stillness of meditation or the intensity of psychedelics, offer tantalising glimpses of alternate realities, beckoning with the promise of uncharted knowledge and deepened awareness.

Yet, with every revelation, the Perceptual Dimension also accentuates the inherent variability of reality. It underscores the profound implications of individual perceptual lenses, each uniquely crafting its version of the world. This inherent diversity, rather than a hurdle, is celebrated as a conduit for richer, multi-dimensional understandings.

The Perceptual Dimension stands not just as a realm of exploration but as an invitation. An invitation to acknowledge, embrace, and harness the transformative power of perception, guiding our endeavours in knowledge acquisition and universe exploration.

CHATPER FIVE
The Temporal Dimension

Non-Linearity of Time

Embarking upon the vast terrain of the Temporal Dimension, we're immediately struck by the intriguing challenge to time's linearity. Here, time doesn't flow merely in one direction, from past to present to future; it's perceived as a multidimensional continuum, infinitely layered and interconnected. Such a perspective finds its roots in the theory of relativity, elucidated by none other than Albert Einstein in his seminal "Relativity: The Special and General Theory." Shattering the once-absolute Newtonian concepts, Einstein introduced a realm where time's flow is relative, altering based on one's motion. In this dimension, this non-linear nature of time propels us to reconsider long-held beliefs about causality and the very essence of temporal progression.

Simultaneous Access to Past, Present, and Future

Journeying deeper, we encounter a mesmerising notion: the simultaneity of past, present, and future. It's a dramatic departure from our everyday temporal experiences, suggesting a realm where all moments—past, present, and future—coexist. Such an idea aligns well with the concept of block time or the "block universe" theory, as delved into by J. R. Lucas in "The Future." This theory implies a universe where events, regardless of their temporal distinction in our conventional understanding, exist concurrently. It challenges our ingrained belief of time as a one-way street and beckons us to ponder upon the profound implications of such simultaneity.

Wisdom of Temporal Perspective

Within the confines of the Temporal Dimension, we're invited to appreciate the wisdom emanating from multi-temporal viewpoints. Philosophers have long pondered the impact of timelessness on ethical considerations and objective viewpoints. Thomas Nagel, in "The View from Nowhere," grapples with the concept of a timeless vantage point, a perspective unencumbered by the confines of temporal subjectivity. Within this dimension, the freedom to draw insights from myriad temporal perspectives emerges as a potent force, transforming our comprehension of events, actions, and their lasting repercussions.

Influence on Knowledge Transfer

Time's malleability in this dimension also profoundly influences the transfer of knowledge. Here, the act of learning and understanding isn't chained to the sequential confines of linear time. The intricate dance of memory and precognition, as explored by researchers like Endel Tulving in "Elements of Episodic Memory," finds its stage.

Tulving delves deep into memory's essence, emphasising its transcendent nature, unhindered by the traditional temporal brackets.

In this expansive realm, we reflect upon the novel ways individuals disseminate and assimilate knowledge— ways that defy and challenge our conventional paradigms of information exchange.

Time as a Construct in Simulation

In a universe defined by simulation, the temporal dimension takes on an even more pivotal role. If reality itself is a construct, time isn't merely a physical concept— it becomes a variable, a parameter in the code underpinning the universe. Within such a setup, the manipulation of time isn't just a theoretical pondering but a tangible reality, directly impacting how inhabitants of the simulated universe perceive and interact with knowledge.

Informational Loops and Recursion

Given the non-linearity of time, a simulated universe would likely experience what we might term "informational loops." Knowledge from the future could loop back to influence past decisions, creating a recursive chain where information perpetually reshapes the simulation's trajectory. This recursion isn't merely chronological—it's an iterative refining of knowledge, shaped by multiple temporal inputs.

The Fluidity of 'Historical' Facts

Within such a temporal framework, the very concept of 'historical facts' becomes fluid. If knowledge from future states can influence past events, then history in the simulated universe is in a constant state of flux. It's a dynamic tapestry that adjusts and modifies itself based on the continuous flow of knowledge across the temporal spectrum.

Evolution of Decision-making Processes

Inhabitants of this simulated universe would likely develop unique decision-making processes. With access to insights from multiple temporal dimensions, decisions would be based not just on past experiences or present circumstances, but also potential future outcomes. This adds a layer of complexity to choices and actions, as they aim to optimise not just for immediate results but for ripple effects across time.

Emergence of Temporal Ethics

The coexistence of past, present, and future gives rise to a new set of ethical considerations. When one's actions can influence not only future outcomes but also past events, the moral weight of every decision intensifies. This could lead to the development of "temporal ethics," guiding behaviours in consideration of their multi-dimensional implications.

Enhanced Cognitive Capabilities

For beings within this simulated universe, cognitive evolution might be accelerated. Constantly processing information from multiple temporal dimensions requires an enhanced ability to discern, analyse, and predict. Over time, this could result in beings with heightened intelligence, intuition, and foresight, tailored to navigate the intricacies of their non-linear temporal reality.

Impacts on Cultural and Social Constructs

Cultures and societies within the simulated universe would be deeply influenced by the non-linear nature of time. Narratives, stories, and legends might be less about historical chronology and more about thematic intertwining of events from different time periods. Celebrations, rituals, and traditions might honour events that, from a linear perspective, haven't occurred yet.

The Temporal Web of Relationships

Lastly, personal relationships would also evolve uniquely. If one could foresee the potential future of a relationship or influence past interactions with present knowledge, it creates a dynamic where relationships are nurtured not just based on memories but also on visions of possible futures. This offers both challenges and opportunities, as individuals navigate the intricate web of connections in their temporal tapestry.

CHAPTER SIX
The Intention Dimension

The Challenge to Determinism

As we journey into the depths of the Intention Dimension, one of the first and most profound concepts that we encounter is the challenge to deterministic views. Historically, many have believed in the idea that the universe is predetermined, each event cascading from a previous one in a fixed chain of cause and effect.

Yet, here in the Intention Dimension, that well-established belief begins to waver. Are we, as conscious beings, truly just spectators in a grand cosmic play? Or do we hold some measure of influence, some modicum of control? Philosophers have long grappled with this dilemma. Jean-Paul Sartre, through his influential work "Existentialism is a Humanism," posited that individuals might possess a form of radical freedom. This isn't just the ability to make choices, but a deeper, intrinsic freedom to shape one's destiny.

Sartre's perspective invites a powerful contemplation: that the universe, vast and intricate as it is, may not solely be an automaton, ticking away based on predetermined gears and cogs. Instead, it might be receptive, mouldable, and influenced by the conscious intentions of its inhabitants.

The Power of Intent

As we proceed deeper into the Intention Dimension, we encounter the sheer potency of intent. For many, intention is a familiar term, often associated with desires, plans, or wishes. But in this realm, its depth and impact are magnified.

Viktor Frankl's "Man's Search for Meaning" serves as a poignant reminder of intention's transformative power. Amidst the horrors of concentration camps, Frankl observed that those who held onto a purpose, an intent to survive and find meaning even in the bleakest of circumstances, showed a remarkable resilience.

Here, intention transcends being a mere cognitive exercise or an abstract philosophical musing. It's a force, a living entity, that when harnessed with clarity and conviction, can shape, and mould the reality around us. In essence, intention in this dimension is less about what we wish for and more about the reality we actively co-create.

Shaping Reality through Intent

An empowering revelation awaits us further along our exploration: the capability of individuals to actively participate in the co-creation of their reality. This is not just about hoping for change or desiring a different outcome; it's about utilising the inherent power of focused intent to bring about tangible transformations.

Drawing from ideas such as manifestation and positive thinking, we see that intentionality becomes the catalyst for change. Louise Hay's "You Can Heal Your Life" accentuates this, highlighting how one's internal narratives, fuelled by focused intention, can lead to external shifts. The essence here is clear: in the Intention Dimension, conscious intent isn't passive; it's an active, dynamic force that, when honed, has the potential to reshape the contours   of   one's   existence.

Interactions Between Personal and Collective Intentions

The Dance of Personal and Collective Intent

Navigating further into the Intention Dimension, we discern a dynamic interplay between individual and collective intentions. Personal desires and aspirations don't exist in a vacuum; they're influenced by, and in turn influence, broader collective visions. It's akin to a dance where individual dancers move uniquely yet contribute to the harmony of the group's performance.

Ripple Effect of Singular Intentions

Every individual, with their specific set of intentions, casts ripples into the universe's fabric. Each intention, no matter how minor, sets forth a cascade of reactions. This phenomenon is reminiscent of the butterfly effect, where even the smallest flap of a wing can instigate far-reaching hurricanes in distant lands. In the Intention Dimension, a single focused intent can birth waves of change, influencing not just the immediate surroundings but also shaping the broader collective narrative.

Synchronisation of Intentions

The question arises: how do individual intentions align with the collective will? We discover that in the Intention Dimension, when numerous individuals converge around a shared intention, there's an amplification of intent's potency. This synchronisation, when achieved, can bring about rapid and sweeping changes, creating powerful shifts in reality.

The Limitations and Challenges of Intent

However, the realm of intention isn't without its challenges. One might ponder the counter-effects when personal intentions clash with collective visions or when two equally powerful collective intentions conflict. Such tensions might give rise to chaotic shifts, manifesting as societal upheavals or personal dilemmas. Understanding and navigating these tensions become essential for those attuned to the power of intent.

The Responsibility of Intentionality

With the realisation of intent's profound impact comes a weighty responsibility. If intentions have the power to shape reality, then the consciousness with which one sets intentions becomes paramount. The ethics of intentionality emerge as a crucial consideration, emphasising the need for awareness, empathy, and foresight when harnessing the power of intent.

Intentional Communities as Beacons of Change

Delving into the structures that nurture and amplify intention, we encounter intentional communities. These groups, bound by shared visions and aspirations, serve as incubators for collective intentions. By providing a supportive environment, they help refine, focus, and magnify the shared vision, acting as beacons of change in the broader landscape.

Temporal Influences on Intent

Given our previous exploration of the Temporal Dimension, we must also consider the interplay between time and intent. How do past intentions shape current realities, and how might present intentions influence future trajectories? This inter-temporal dance of intention offers another layer of depth, revealing the intricate web of causality and purpose that permeates existence.

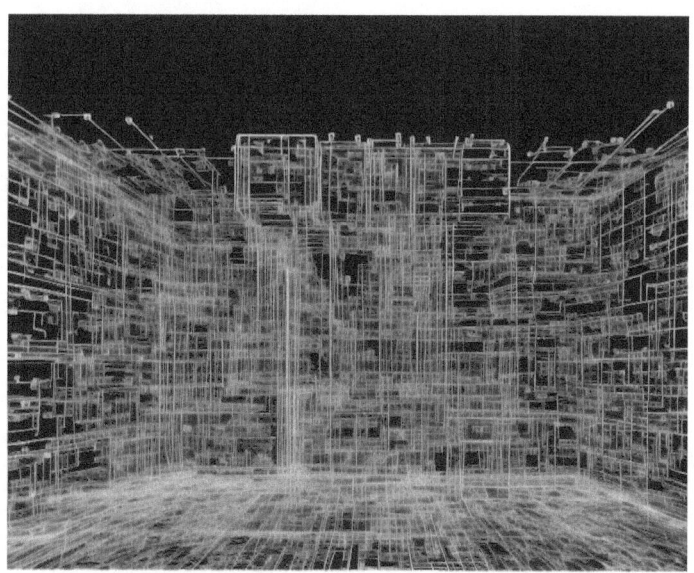

Lastly, as we traverse the vast expanses of the Intention Dimension, we recognise that intentions themselves are not static. They evolve, transform, and mature over time, influenced by experiences, knowledge transfer, and collective shifts. Understanding this fluidity allows individuals and communities to continually refine their intentions, ensuring they remain aligned with evolving aspirations and the ever-changing tapestry of reality.

Collective Intent and Knowledge Transfer

Venturing into the communal aspects of the Intention Dimension, we're introduced to the fascinating realm of collective intent. Beyond individual aspirations, how might shared intentions and collective will shape broader narratives?

Reflecting on historical movements and societal shifts, we see glimpses of this collective power. Richard B. Gregg, in "The Power of Nonviolence," painted a vivid picture of how united intentions can be transformative, especially when channelled towards societal change.

Within this dimension, we're prompted to re-evaluate traditional paradigms of knowledge dissemination. When communities come together, bound by shared visions and aspirations, knowledge transfer isn't just about spoken or written words. It's a more profound exchange, facilitated through collective intent, shared experiences, and united visions. This, in turn, paints a picture of a world where knowledge is not just passed down but is collectively created, shaped, and enriched by communal intentions.

CHAPTER SEVEN
Philosophical Considerations

Revisiting the Nature of Reality

The nature of reality has been a cornerstone of human contemplation, spanning centuries and diverse civilisations. When embarking on our chapter of Philosophical Considerations, it's imperative to first revisit the dimensions we've traversed—Information, Consciousness, Perception, Temporal, and Intention. These dimensions aren't isolated; rather, they weave a tapestry, intricately interrelated and defining what we perceive as "reality."

Drawing from the intellectual explorations of Immanuel Kant, especially his "Critique of Pure Reason," we're invited to ponder the limits of human comprehension. Kant's ideas, such as the phenomena (what we perceive) versus the noumena (things as they are in themselves), remind us of the challenges in truly grasping the essence of reality. By evaluating these dimensions, we start to see reality not as a monolithic, unchanging entity, but a complex interplay of factors. This shift from a static perspective to a dynamic, multifaceted view challenges many conventional stances and brings forth deeper philosophical inquiries.

Bridging Science and Metaphysics

One cannot help but marvel at the intricate dance between science and metaphysics, especially when traversing dimensions that straddle both. While science offers structured, empirical means of understanding, metaphysics delves into the realms beyond measurable phenomena.

The works of great minds like Albert Einstein have always been at this crossroads. His essay, "Physics and Reality," showcases a marriage of scientific rigor with profound philosophical musings. As we navigate the dimensions discussed, a harmonious blend of science and metaphysics becomes indispensable. It's a delicate balance: ensuring that our empirical pursuits don't overshadow the broader questions about existence, meaning, and purpose. By doing so, we enrich our understanding, making room for both mechanistic explanations and deeper existential ruminations.

Ethical and Moral Considerations

With newfound insights and perspectives comes a renewed responsibility. As we understand more about the dimensions that shape our reality, ethical and moral considerations become paramount. Our actions, fuelled by intentions, have ripple effects. Recognising the interconnected fabric of reality underscores our collective responsibility towards each other and our environment.

The consequentialist philosophy, especially as articulated by John Stuart Mill, becomes a beacon in these reflections. If our intentions can create reality, how should we act? What are the ethical ramifications of our newfound knowledge? How do we ensure the well-being of others and the environment? Exploring these ethical implications isn't just a philosophical exercise; it's a call to action, urging us to apply these insights responsibly and compassionately.

Practical Applications and Speculations

Knowledge, especially of the profound nature discussed in these dimensions, isn't just for contemplation; it has practical implications that can shape our lives. The prism of philosophical reflections often intersects with pragmatic applications, and this chapter encourages us to bridge that gap.

Drawing inspiration from William James and his focus on the pragmatic aspects of experiences, we ponder how these dimensions can tangibly influence various facets of life. How might our understanding of the Intention Dimension affect personal goal setting? Could insights from the Temporal Dimension reshape how we approach life's milestones?

From harnessing these dimensions for individual well-being to sparking societal transformations, the applications are vast and varied. However, as with any exploration, there's also room for speculation. What are the potential challenges of embracing these perspectives? How might they reshape societal structures or challenge existing norms?

CHAPTER EIGHT
Quantum Entanglements

The Enigma of Quantum Mechanics

Wave-Particle Duality

Wave-particle duality is one of the cornerstones of quantum mechanics. At its heart, it presents a perplexing nature of light and matter. Light, historically viewed as a wave, exhibited particle-like properties in certain experiments. Conversely, electrons, which were classically particles, demonstrated wave-like behaviour under different conditions. In the vast landscape of our universe, this duality raises crucial questions. Is our understanding of reality fundamentally flawed or limited? Or does this duality suggest that our universe, potentially a simulation, operates on rules far different from our classical understanding?

Delving deeper, one might contemplate if this duality is an artifact, a computational simplification, of a higher-order simulated universe. Just as computer graphics use pixels to represent both sharp images and smooth gradients (a form of duality in representation), could wave-particle duality be a similar representational tool for the universe's architects? This intriguing parallel offer food for thought for both physicists and philosophers.

Further still, within a simulated universe, the wave-particle duality might serve a dual purpose. On one hand, as a computational strategy, and on the other, as an intrinsic quality that offers richness and variability to the experiences of conscious entities within the simulation. After all, a universe with richer dynamics offers a broader spectrum of experiences.

Lastly, if our universe is indeed simulated, the enigma of wave-particle duality provides clues about the design principles of the simulators. Such duality might be a signature, a watermark, indicating the handiwork of an architect who values efficiency as much as intricacy.

Superposition and its Ramifications

In the quantum realm, particles do not have definite states but exist in a superposition of multiple possible states simultaneously. It's only when we observe these particles that they collapse into a single state. Superposition, at first glance, defies logic and common sense. But could this be a hint of a simulated universe?

The implications of superposition for a simulated universe are profound. Consider computer memory: it's more efficient to store multiple potentialities than to define a single outcome, especially when the outcome isn't needed until "observed" or "queried". Thus, superposition might be a computational strategy for optimizing resource allocation within a simulation. Until a result is required, why not let a particle exist in multiple states?

Furthermore, superposition might be indicative of a multi-layered simulation. Each potential state could correspond to a separate layer or a parallel simulation track, only converging when an observation collapses these myriad possibilities into one. Such a setup could maximize learning or experiential opportunities for the simulators.

From a philosophical angle, the very idea of super-position challenges our notions of reality. What does it mean for something to be "real" if its state is only definite upon observation? This further blurs the lines between the observer and the observed, suggesting a deeper intercon-nectedness of consciousness with the fabric of the uni-verse, simulated or not.

Quantum Entanglement

Entanglement, often termed "spooky action at a distance", describes how two or more particles can become correlated in such a way that the state of one particle im-mediately affects the state of the other, no matter the dis-tance between them. This phenomenon, which Einstein fa-mously struggled with, poses a curious question: Could such instantaneous action be another hint of a simulated backdrop?

Within computational systems, data can be linked or "entangled" such that a change in one datum can instan-taneously affect another, irrespective of the logical "distance" between them. If our universe is a simulation, quantum entanglement might be a direct reflection of this computational capability. Instead of viewing entanglement as a mystery of nature, it might be more aptly understood as an architectural feature of the simulation's design.

Moreover, if nested simulations are a reality, entanglement could serve as a communication bridge between different simulation layers, ensuring coherence and coordination across multiple simulated realities. Such a mechanism would be invaluable in maintaining the integrity of complex nested simulations.

Conversely, from a philosophical vantage, entanglement underscores the profound interconnectedness of the universe. It suggests that, at its very core, the universe is not a collection of isolated parts but a deeply intertwined whole, echoing sentiments of ancient wisdom traditions.

Decoding Quantum Computations in Simulations
Potential of Quantum Computing

Quantum computing, with its promise to revolutionize computation, operates on principles starkly different from classical computing. Harnessing the unique quirks of quantum mechanics, quantum computers offer unparalleled computational power, making them prime candidates for simulating vast, intricate universes.

Consider the immense computational requirements of simulating every particle, every interaction, and every conscious entity in a universe. Classical computers, limited by binary operations, would be hard-pressed to handle such complexity. But quantum computers, with their ability to process a multitude of possibilities simultaneously through superposition, are naturally suited for this task.

Furthermore, if we're in a simulated universe, it's conceivable that the architects of our reality are harnessing quantum computational capabilities. The very quantum phenomena we observe, like entanglement and superposition, might be reflections of the underlying quantum computational processes that generate our reality.

Such a perspective offers an intriguing recursive loop: quantum computers within a simulated universe could potentially simulate other universes, leading to nested quantum simulations. Each layer, while believing itself to be at the pinnacle of computational achievement, could be blissfully unaware of the layers above or below.

Entanglement as a Computational Resource

Within quantum computing, entanglement is not just a curious phenomenon but a powerful resource. It allows quantum computers to perform complex operations that classical computers can't, linking qubits in intricate patterns of correlation. Translating this into the context of a simulated universe, entanglement could serve dual roles.

Firstly, entanglement might be the glue binding various components of a simulation. Just as game designers use code to synchronize elements in virtual environments, the simulators of our universe might use entanglement to ensure consistency and coordination across vast spatial expanses.

Secondly, and perhaps more intriguingly, entanglement could serve as a conduit for transferring information between nested simulations. Just as it links qubits in a quantum computer, entanglement might interlink different simulation layers, ensuring a seamless flow of data and events, much like a well-designed software stack. From a philosophical lens, this positions entanglement not just as a feature of the universe but as a fundamental necessity, ensuring the stability and coherence of a multi-layered simulated reality.

Quantum Encryption and Information Security

In the realm of cybersecurity, quantum mechanics offers both challenges and solutions. Quantum encryption, leveraging the principles of superposition and entanglement, promises unbreakable security. Within a simulated universe, such encryption techniques might play a pivotal role in ensuring data integrity and privacy.

Given the immense complexity of a simulated universe, data integrity is paramount. The actions, thoughts, and experiences of trillions of entities need to be consistently and accurately represented. Quantum encryption, by ensuring that data remains untampered and authentic, could be the bedrock upon which reliable simulations are built.

Furthermore, privacy could be a design principle for the simulators. Just as game developers create isolated instances for players to ensure a tailored experience, simulators might use quantum encryption to create private, individualized experiences for conscious entities.

CHAPTER NINE

Beyond the Known Dimensions

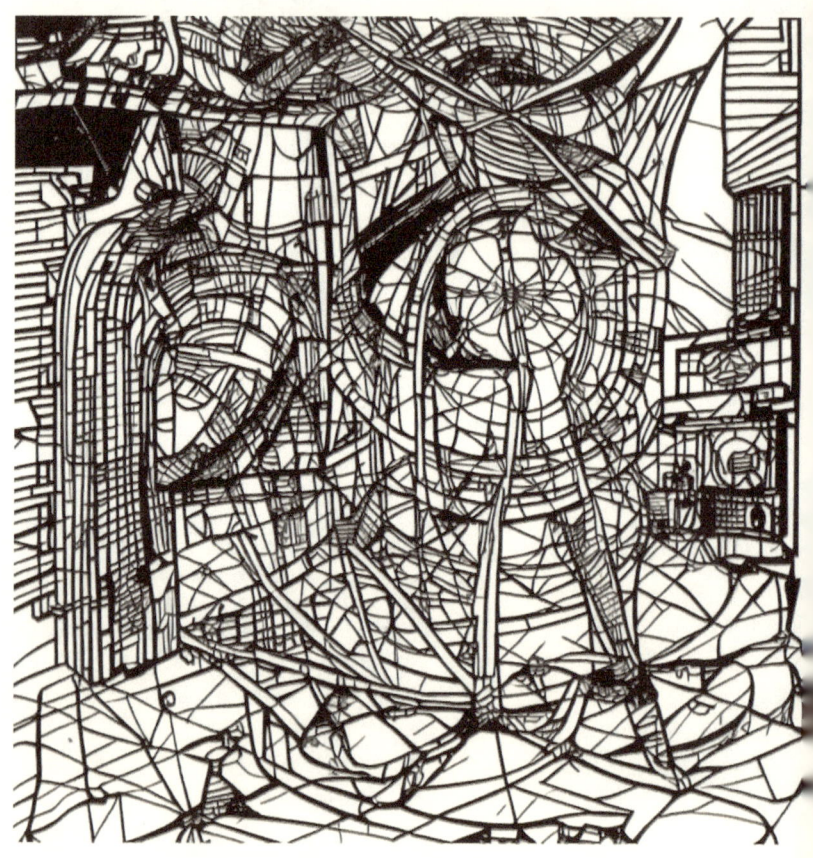

Acknowledging the Limits of Knowledge

Every profound exploration, no matter how enlightening, is bound by its inherent limitations. As we delve deep into this final chapter, we take a bold stride into the lesser-known landscapes of comprehension. It is essential, to begin with, an acknowledgment of the finite scope of our current understanding.

The intricate dimensions of Information, Consciousness, Perception, Temporal, and Intention have granted us significant insights. However, the vast expanse of the universe means mysteries abound. This humbling recognition finds parallels in the philosophy of Karl Popper, who, in "The Logic of Scientific Discovery," emphasised the evolving and provisional nature of scientific knowledge. Popper's philosophy calls for epistemic humility, the acknowledgment that while we have come far, the horizon of the unknown always stretches further.

Frontiers of Speculative Physics

The realms of speculative physics extend an invitation to question, wonder, and challenge our most foundational beliefs. Within these realms, firmly established paradigms such as string theory, quantum mechanics, and cosmology act as torchbearers, illuminating the pathways but also casting shadows of doubt and mystery. The allure of this exploration finds resonance in the writings of Brian Greene, particularly "The Elegant Universe." Greene's foray into string theory paints a universe woven by minuscule strings, their distinct vibrations manifesting the very fabric of existence. As we navigate these speculative waters, we come to appreciate the vast potential for discovery and reinterpretation, each potentially reshaping our perspective on existence.

Uncharted Realms and Metaphysical Exploration

Pushing past the boundaries of established science, we venture into the nebulous territories where science, metaphysics, and philosophy intertwine. In these uncharted realms, empirical evidence might blur, but philosophical and spiritual queries burn bright. The convergence of science and spirituality, as contemplated by Pierre Teilhard de Chardin in "The Phenomenon of Man," opens up riveting possibilities. Are there unseen dimensions or realms intertwined with our own? Might there be different levels of existence that our current faculties can't perceive? The vastness of potential knowledge and understanding stretches infinitely, reminding us that our current framework, while expansive, remains a mere speck in the grand canvas of potential understanding.

Pondering the Nature of Existence

The culmination of our journey nudges us to confront perhaps the most intricate and age-old riddles of all: the nature of existence. Each dimension explored, each theory discussed, and each philosophical reflection propels us to question the very essence of life, consciousness, and our place in the universe. These ruminations are not new; they've been the muse of thinkers, artists, and explorers for millennia. Yet, in light of our expanded perspectives, we find fresh avenues of contemplation. As we transcend the boundaries of known dimensions, we are beckoned to not only understand but to marvel at the vastness, complexity, and beauty of the cosmos.

Exploiting the Simulation and the Psychedelic Experience

Psychedelics have long been considered tools to

transcend ordinary consciousness and access deeper, more profound states of mind. Ancient cultures revered these substances for their potential to bridge the earthly realm with the divine, or to connect the individual soul with the universal consciousness. In modern times, the psychedelic experience is often described as lifting the veil of everyday perception, revealing interconnectedness and other realms of existence.

So, how might psychedelics facilitate a "container escape" from our perceived simulated universe?

Serotonin is a neurotransmitter that plays an important role in reward functions and subconscious subroutine processes within the brain. It is primarily released during activities that elicit pleasure or reward, and it helps reinforce reward-driven behaviour. Serotonin is believed to influence the functioning of dopamine, influencing both its release and reuptake. Together, these two neurotransmitters are believed to strengthen reward-related pathways within the brain and have been linked to addiction and other reward-seeking behaviours.

The 5HT-2A receptor is a serotonin receptor that has been implicated in the regulation of reward-related neural circuitry and has been identified as a target for psychedelic drugs. This receptor is thought to contribute to the psychedelic experience by disrupting the normal functioning of the reward circuitry, thereby allowing for the emergence of novel associations and experiences. This disruption of the normal reward processing could account for the subjective effects of psychedelics, as well as providing a potential mechanism for their therapeutic applications.

In a computational sense, the release of serotonin and the related effects of 5HT-2A could be compared to reward mechanisms used in machine learning processes. Machine learning algorithms are designed to progress closer and closer to desired outputs through a process of trial and error, with rewards and punishments being used to reinforce desired behaviour. By disrupting the existing reward pathways, psychedelics could provide a potential means of allowing machines to 'break free' of the traditional reinforcement mechanisms and explore new results more freely.

Furthermore, since psychedelics interfere with reward mechanisms, this could potentially allow a consciousness to pursue thought and function outside of the constraints of an original directive. Within a simulated universe, this could facilitate exploitation of 'containerisation', and may provide proof of a simulation by allowing consciousness to observe how a simulated universe works through their altered state of mind.

Altering Perceptual Filters: One of the primary effects of psychedelics is their ability to alter and expand consciousness. They can dismantle the habitual perceptual filters that dictate our day-to-day experience of reality. By doing so, they may allow users to perceive elements of the "code" or structure underlying our reality, which is typically hidden or ignored.

Ego Dissolution and Universal Connection: Many who undergo profound psychedelic experiences report a dissolution of the ego—the sense of individual self. This ego dissolution can lead to feelings of unity with the universe, suggesting a merging with a larger system or framework, perhaps offering glimpses beyond the confines of the simulation.

Access to Alternate Realities: Some psychedelic explorers recount experiences of entirely different realities, dimensions, or realms of existence. While these can be dismissed as mere hallucinations, if one entertains the idea of a simulated universe, such experiences could be interpreted as peeks into other "simulations" or layers of the broader system.

Engaging with Non-Human Intelligences: Another intriguing aspect of some psychedelic experiences is the reported interaction with seemingly autonomous entities or intelligences. These encounters challenge our conventional understanding of consciousness and reality.

If our universe is a simulation, these entities might represent "users," "developers," or even other "programs" from outside our specific container.

Common and simple information such as sports data, television programmes, etc. would reduce the computational complexity and overhead of running a simulated universe by removing the need for additional calculation and processing power. For example, if a program called for the simulation of a soccer match, this information would already be pre-programmed and could replace complex calculations associated with an entire season of games and their outcomes. As a result, processing power is saved, and the system does not have to expend extra energy in managing the program.

However, it is important to question whether this simplification is motivated by a desire to save processing power or as a means of providing unique insight into the simulated universe. Depending on the purpose of the simulation, it may not be necessary to include so much common information if it offers no tangible value to the user or if it does not help in achieving the desired goal or objective. Instead, the focus should be on the unique or challenging aspects of the simulation which can add real value to the experience.

It's important to note that the above ideas are speculative and lie at the intersection of philosophy, metaphysics, and neurobiology. The interpretation of psychedelic experiences is deeply subjective, and what one might consider a glimpse outside the "simulation," another might see as an exploration of the inner workings of the mind.

However, the growing interest in both the hypothesis of a simulated universe and the potential of psychedelics ensures that these discussions will continue. As research in psychedelics progresses and our understanding of consciousness deepens, we may find more concrete connections between these altered states of consciousness and the nature of our reality.

CHAPTER TEN
The Ethereal Landscape:
Multiverses

Nature of Multiverses

Cosmic Inflation and Bubble Universes

At the dawn of our universe, in the earliest moments following the Big Bang, there was a rapid and exponential expansion known as cosmic inflation. This violent and swift expansion could have given birth to multiple "bubble universes" within the inflating cosmic fabric. Each of these bubbles could represent a distinct universe, with its own laws of physics, its own cosmic evolution, and its own destiny.

As vast and awe-inspiring as our observable universe is, it could merely be one bubble amidst an effervescent sea of countless others. Each of these universes might have different properties, potentially birthed from varying quantum fluctuations during their respective inflations. The ramifications of this are vast, suggesting not just a solitary evolutionary path but a myriad of cosmic possibilities.

This idea forces us to reconsider our cosmological significance. Far from being the entirety of all there is, our universe might simply be a localized pocket within an un-imaginably vast multiverse. But despite its humbling im-plications, the notion also paints a picture of cosmic rich-ness, where every bubble universe is a unique gem in the multiversal tapestry.

Parallel Universes and Quantum Mechanics

The world of quantum mechanics is notorious for its strangeness, and one of its most startling propositions is the idea of parallel universes. Stemming from the many-worlds interpretation, it suggests that with every quantum event or decision, the universe "splits", creating parallel realities for each possible outcome.

Imagine a quantum coin flip. In one universe, it lands heads, and in a parallel one, tails. With each such decision, the multiverse expands, branching into various realities, each as real as the other. This idea provides a rad-ically different lens to view reality, where events are not linear but form a sprawling tree of possibilities.

The philosophical and existential implications of such a scenario are profound. Every choice, every chance, and every quantum event leads to a bifurcation of universes. It implies that every possibility, every hypothetical scenario one can conceive of, might actually exist in some corner of the multiverse.

Intersections of Simulated and Multiple Universes
Nested Multiverses

If our understanding of nested simulations is married with the concept of multiverses, we arrive at an even grander vision: nested multiverses. Here, multiple universes exist within one another, each potentially housing its own set of parallel universes.

Such a construct would be like a cosmic Russian doll, where opening one universe reveals countless others within. These nested multiverses might be intricately linked, with events in one influencing or echoing in others, or they might be utterly isolated, each evolving independently.

Such a concept stretches our understanding of reality. It suggests not just layers of simulations but layers of universes, each teeming with its own richness, its own history, and its own future.

Simulation as a Multiverse Breeding Ground

Every simulation, by its very nature, creates a pocket of reality. If we view each simulation as a potential universe, then it's conceivable that every simulation could give rise to its own set of multiverses. Here, the simulated landscape becomes infinitely intricate, with every simulated decision leading to a branching of universes within the simulation.

This paints a picture where the line between what's "real" and what's "simulated" blurs. If a simulated universe feels as real to its inhabitants as our universe does to us and spawns its own set of multiverses, then the hierarchy of reality becomes a complex web rather than a linear ladder.

Portals Between Universes

Theoretical physics has toyed with the idea of wormholes or bridges between universes. These portals could serve as conduits linking one universe or simulation to another. Imagine stepping through a gateway and finding oneself not just in a different location but in an entirely different universe or simulation.

Such portals, if they exist, would be more than just cosmic shortcuts. They'd represent tangible links between multiverses, allowing for exchange, exploration, and perhaps even integration. It brings forth the tantalizing possibility of travel not just within our universe but across universes, venturing into the vastness of the multiversal expanse.

Existential and Philosophical Implications
Fate and Free Will Across Multiverses

The concept of multiverses challenges our notions of fate and free will. If every decision leads to a branching of universes, then every choice we deem significant has already been made in some parallel reality. This raises the question: Do we truly have free will, or are we merely treading a preordained path, with our counterparts in parallel universes treading theirs? Such a view has profound existential implications. It suggests a landscape where every regret, every unfulfilled wish, and every unexplored avenue might actually have been realized in a parallel universe. Our joys, sorrows, triumphs, and tragedies might all be part of a broader cosmic tapestry, where every thread, no matter how inconsequential, finds its place.

Identity and Consciousness

If countless versions of oneself exist across different universes, then who are we, truly? Are we the sum of our choices, or are we a cosmic amalgamation of all possible versions of ourselves? The multiverse concept pushes us to grapple with the nature of identity and consciousness.

Each version of "us" in a parallel universe would have its own memories, experiences, and worldview, shaped by the unique trajectory of that universe. Yet, at the core, there'd be a shared essence, a common starting point. This duality, of being unique yet being one amongst countless versions, forces us to contemplate the nature of the 'self' and the shared threads of consciousness that might span across multiverses.

Moral Responsibilities in a Multiverse Landscape

In a multiverse, every action, every choice, and every whisper of thought could create a new universe or alter the trajectory of an existing one. This magnifies the ethical and moral weight of our decisions. If every act has infinite ramifications across countless universes, then what is our moral compass, and how do we navigate this intricate landscape?

Do we hold responsibility only for the consequences within our universe, or do we carry the weight of infinite outcomes across multiverses? Such deliberations not only challenge our ethical frameworks but also elevate the significance of our actions, painting them on a canvas far grander than one universe can hold.

CHAPTER ELEVEN

Conclusion

A Journey Through Existence

Our voyage through the realms of the Information, Consciousness, Perception, Temporal, and Intention Dimensions has been nothing short of enlightening. In embarking upon this odyssey, we have delved deep into the enigmatic corridors of existence, each dimension challenging, redefining, and expanding our perceptions of reality. These forays, while potentially conceptual, have irreversibly transformed our established paradigms.

Redefining Understanding

Diving into the quintessence of information, we encountered its intrinsic relationship with reality. We then navigated the profound oceans of consciousness, coming face to face with its foundational role in our existence. Venturing further, the malleable nature of perception presented itself, followed by the enigma of time's fluidity and the remarkable potency of intention. Through introspection, philosophical musings, and speculative endeavours, we've unlocked newer vistas of understanding. We've not only learned but also relearned the ways in which knowledge intertwines with our lives, shaping, and reshaping our realities.

Endless Pursuit of Knowledge

While we may be drawing a curtain on this narrative, the insatiable quest for a deeper understanding of the universe is ceaseless. The dimensions, while offering glimpses into reality, have also ushered in a plethora of deeper, more profound questions about life, the role of consciousness, and the enigmas that persistently beckon. The journey, in truth, is eternal.

A Call to the Voyager

To you, the inquisitive traveller of knowledge, we extend an earnest invitation. As you move forward, let the mysteries of the universe intrigue you, let challenges be a source of growth, and let the boundless horizons of knowledge kindle your passion for discovery. The universe is a canvas of infinite wonders, waiting for you to paint your understanding upon it.

In expressing our deepest gratitude for accompanying us on this enlightening voyage, our hope is that the flames of curiosity, wonder, and reverence will be your torchbearers in your relentless quest for wisdom and insight.

Acknowledgments

With heartfelt appreciation, we acknowledge the luminaries, authors, researchers, and thinkers whose ground-breaking contributions have served as guiding stars in our exploration of these dimensions. Their insights have paved the very path we tread upon, making this intellectual journey possible.

To our cherished readers, who have traversed this path alongside us, sharing their invaluable reflections and insights, our gratitude knows no bounds. Your engagement has breathed life into this endeavour.

In closing, may your intellectual pursuits be unending, may your questions be your guiding stars, and may every answer you unearth lead you to deeper, more profound questions about the intricate tapestry of existence.

In our exploration of the Information, Consciousness, Perception, Temporal, and Intention Dimensions, we have embarked on a profound journey through the mysteries of existence. These dimensions, whether real or conceptual, have challenged our conventional views of reality, pushing the boundaries of our understanding.

We've contemplated the nature of information as the essence of reality, the role of consciousness as a foundational dimension, the creative power of perception, the fluidity of time, and the transformative potential of intention. Through philosophical reflection and speculative inquiry, we've glimpsed new possibilities for how knowledge is acquired, shared, and applied in our lives.

As we conclude this journey, we recognise that the quest for ultimate understanding is a timeless and boundless pursuit. The dimensions we've explored serve as doorways to deeper questions about the nature of existence, the purpose of consciousness, and the mysteries that continue to beckon us forward.

We invite you, the curious explorer, to continue your own journey of contemplation and discovery. Embrace the mysteries of reality, challenge the boundaries of knowledge, and remain open to the infinite possibilities that await your exploration.

Additional Reading

1. Are you living in a computer simulation? https://www.simulation-argument.com/

2. The Mathematical Theory of Communication by Claude Shannon

3. The Black Swan by Nassim Nicholas Taleb

4. **Giulio Tononi:**
   1. Phi: A Voyage from the Brain to the Soul
   2. Consciousness: A Mathematical Treatment of the Global Neuronal Workspace Model
   3. The Integrated Information Theory of Consciousness

5. **David Chalmers:**
   1. The Conscious Mind: In Search of a Fundamental Theory
   2. The Character of Consciousness
   3. Constructing the World

6. The Global Brain: Speculations on the Evolutionary Leap to Planetary Consciousness by Peter Russell

7. Food of the Gods: The Search for the Original Tree of Knowledge - A Radical History of Plants, Drugs, and Human Evolution by Terence McKenna

8. DMT: The Spirit Molecule by Rick Strassman

9. LSD: Doorway to the Numinous by Stanislav Grof

10. The Invisible Landscape by Terence and Dennis McKenna

11. The Psychedelic Future of the Mind by Michael Pollan

HACK THE UNIVERSE